Note

This highly condensed booklet has been arrang
easily located. However, concepts necessary fc
read in the order they appear.

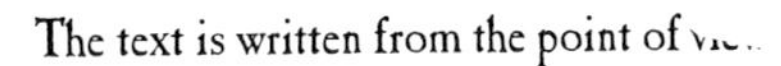
The text is written from the point of vi..

Contents

Find out more about the Moon by visiting

www.moongage.co.uk

(free teacher resources available online)

The Amazing Moongage™

Accurate to within 1 day in 122 years, the Moongage Perpetual Lunar Chart enables you to gauge the whole of the Moon's cycle and the days of its principal phases at a glance, every month, year after year.

Diagram 1.

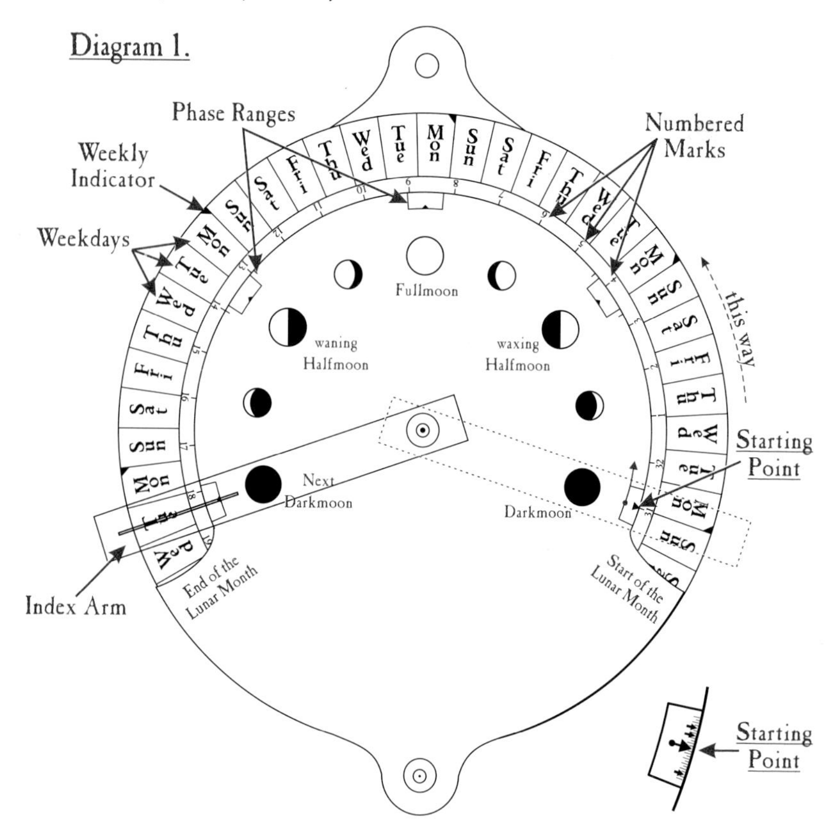

Assembling Your Moongage

It is strongly recommended that you study the diagrams above and opposite, and read through the instructions first before attempting to assemble your Moongage.

You should find 3 card pieces, 1 see-through Arm and 2 Drawing Pins pushed into 2 Fasteners. Carefully turn the Fasteners back and forth to pull out the Drawing Pins.

Diagram 2.

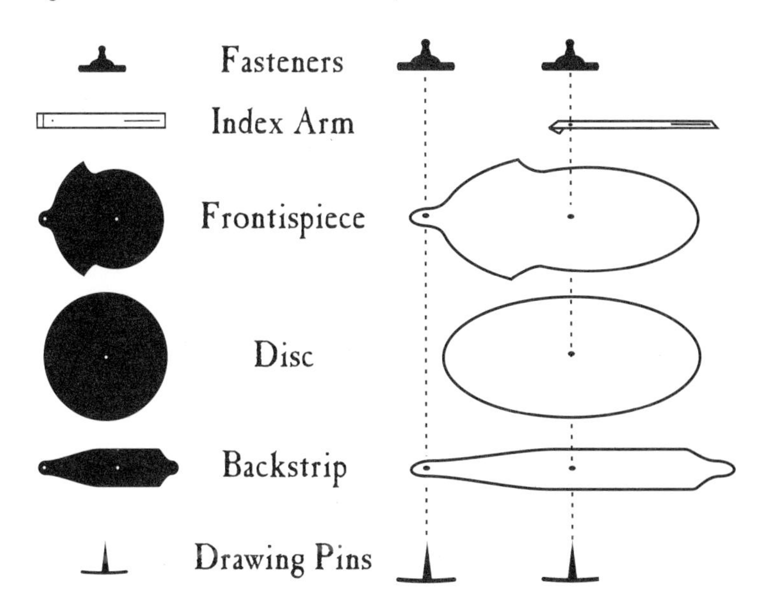

- First use an ordinary pin to pierce holes through each component where indicated by a central black dot or a white dot at one end (Diagram 2)

- Fold a flap under the transparent Index Arm where indicated by the dotted white line

- In the order shown in Diagram 2 assemble all the parts together by pushing the Drawing Pins through the pin-holes from underneath, and then finally affixing the Fasteners back onto the Drawing Pins

Your assembled Moongage should now look like Diagram 1

Setting Your Moongage

You need to set your Moongage only once to keep it in track with the Moon -

- Look for today's date or the nearest previous date in the Setting Table overleaf. Make a note of the red number to the right of it. So if the date is 1st Jan 2004, the nearest previous date on the table is 23 DEC 2003 and its number is **13**

Setting Table

New Moon	●	New Moon	●	New Moon	●
4 DEC 2002	32	25 OCT 2003	11	14 SEP 2004	22
2 JAN 2003	1	23 NOV 2003	12	14 OCT 2004	23
1 FEB 2003	2	23 DEC 2003	13	12 NOV 2004	24
3 MAR 2003	3	21 JAN 2004	14	12 DEC 2004	25
1 APR 2003	4	20 FEB 2004	15	10 JAN 2005	26
1 MAY 2003	5	20 MAR 2004	16	8 FEB 2005	27
31 MAY 2003	6	19 APR 2004	17	10 MAR 2005	28
29 JUN 2003	7	19 MAY 2004	18	8 APR 2005	29
29 JUL 2003	8	17 JUN 2004	19	8 MAY 2005	30
27 AUG 2003	9	17 JUL 2004	20	6 JUN 2005	31
26 SEP 2003	10	16 AUG 2004	21	6 JUL 2005	32

- Line up this same Numbered Mark on the Disc with the Starting Point on the Frontispiece (Diagram 1). These numbers represent New Moons, not dates of the month

- From the Starting Point move the Index Arm *anticlockwise* the number of days since the date you noted from the Setting Table. So if it's 1st Jan 2004 for example, the previous date on the Table being 23 DEC 2003, move the Index Arm *anticlockwise* from the Starting Point by the difference (9 Weekdays) correcting this if necessary by making sure the Index Arm finally lies above today's Weekday.

Using Your Moongage

Now you've Assembled and Set your Moongage using it is easy!

A) Move the Index Arm *anticlockwise* one Weekday every day (when you go to bed or when you wake up in the morning for example - the Moon moves in this direction across the stars). The Index Arm will show you where you are in the Moon's cycle. You can also see which days the Moon's phases will fall upon that month by noting the days that fall within the Phase Ranges (see Diagram 1 & Hints & Tips)

B) When you reach the Next Darkmoon at the End of the Lunar Month turn the Disc forward one number, and then move the Index Arm right back to sit above the same Weekday at the Start of the Lunar Month. So if the Weekday was a Tuesday and the Disc was set to number **31**, you would first move the Disc onto number **32** and then turn the Index Arm back to sit above the Tuesday at the Start of the Lunar Month.

C) Continue repeating the previous 2 steps and the Moongage will keep you in track with the Moon for years to come. It's as easy as ABC!

Hints & Tips

Reading the Phases of the Moon

The Index Arm indicates which stage of the Lunar Month you are in. However, due to the complex nature of the Moon's orbit the length of each Lunar Month (*Lunation*) varies back and forth by up to a day or so each month.

This wandering backwards and forwards from one Lunation to the next produces a range of timings for the Moon's phases.

On the Moongage the range for each principal phase ● ◐ ○ ◑ is indicated by a Phase Range (see page 4)

The Moongage will give you the time of each of the principal phases to the nearest day. For example, the night of the Fullmoon ○ will usually be the midnight indicated by the line between the two Weekdays covered by its Phase Range (e.g. in the diagram on page 4 this would be the Monday night, two weeks after the Darkmoon).

Predicting the Phases More Accurately

The Moongage will help to acquaint you with the Moon's natural cycle. Whilst a standard almanac provides a precise hour and minute for a lunar phase the Moongage maps out the whole experience of the Moon's cycle, providing you with a much wider perspective. As you look out each day throughout the month you will soon discover for yourself the natural behaviour and timing of the Moon. For example -

The Night of the Fullmoon

On the night of the Fullmoon, the Moon reaches its highest point in the sky (its *zenith*) at midnight. At the same time the Moon will be due South (see below).

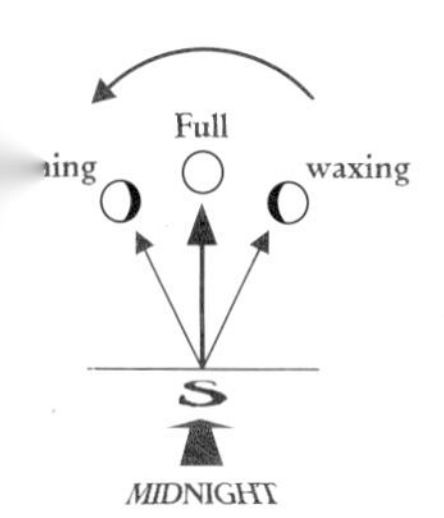

So you can be sure of the night of the Fullmoon by looking where the Moon is in the sky at true midnight. The Moon's relative position in the sky moves in an *anticlockwise* direction by about $12\frac{1}{5}^\circ$ each day. So the day before the night of the Fullmoon the Moon will appear about 12° west of South at midnight. The midnight after the Fullmoon it will appear to be about 12° east of South.

Different Time Zones

The Moongage can be set for any Time Zone by using a different Starting Point to align with the Numbered Marks. The lines above the main arrow (1) represent Time Zones *behind* GMT; the lines below represent Time Zones *ahead*. Each line represents 1 hour difference. The following Time Zones have been marked with arrows. Choose the one nearest to you and use that as your Starting Point -

1 - Great Britain, GMT
2 - Eastern Time, GMT -5
3 - Pacific Time, GMT -8
4 - Canberra, GMT +10

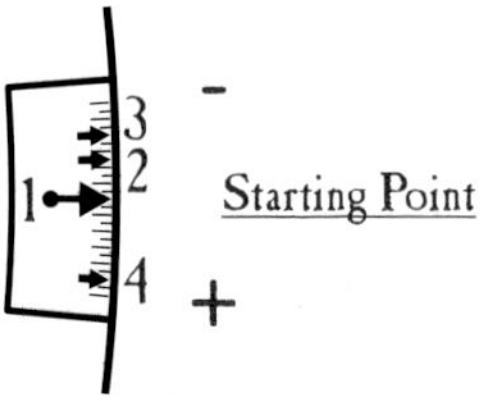

-
3
2
1
Starting Point
4
+

Losing Track of The Moongage

- If you've been away and left your Moongage unattended simply move it forward as many days as you were away, placing it on the correct Weekday
- If you know you're going away then maybe set it ahead of time for when you come back. You can always reset the Moongage from the Table on page 6, or easier still, simply set it from a friend's Moongage!

Maintenance

The Moongage will last many years of constant use when looked after -

- Avoid placing in damp or steamy environments such as a kitchen or bathroom. Damp air causes the boards to warp
- For the same reason the Moongage in its current form is not suitable for prolonged outdoor use
- From time to time press the Fasteners firmly back down onto the Drawing Pins to counter any loosening
- Gently remove any dust from the boards with a clean, dry cloth

The Way of the Moon

The Moon's Story

From time immemorial the Moon has taken centre stage amongst the stars, ceaselessly circling around its partner the Earth. So vital is the Moon that without its presence maintaining the tilt of the Earth's axis, the pulse of the tides and the breeding and hunting cycles of the organisms below there may never have been any life on Earth nor any human civilisation. How has our planetary partner come to influence us so profoundly?

In the Beginning

There is much to suggest that both the Earth and the Moon were born of an immense planetary collision early in the formative stage of the Solar System. The merging bodies of two molten, spinning protoplanets may have formed a very temporary parent-planet which through the force of impact and the centrifugal effect of its spinning body would have rapidly divided to form into a new pair of siblings - our Earth and our Moon. A truly interdependent couple, even though the Moon remains on an imperceptibly slow journey away from its partner [1].

Possibly as a result of being spun away from the surface of the original parent-planet the Moon keeps the same blemished face pointing towards us (as the face of a ball attached to a string remains facing toward us as we spin it around and let it out). These smooth dark blemishes, known as Maria ('seas'), are almost entirely absent from the farside of the Moon. Being much closer to the Earth at a time when there was much more orbiting debris in the Solar System the Earth may have shielded the nearside from many impacts, whereas the farside would have remained more exposed, its Maria blasted out of recognition.

On the other hand the Moon's gravitational influence helps to maintain the Earth's axis at a relatively stable angle. Without the Moon's presence the tilt of the Earth's axis would fluctuate wildly, resulting in highly unstable global temperatures making it impossible for life or even water to settle and gain a foothold without burning, boiling or freezing solid [1].

The proximity of these *binary* planets early on in their history will have made for intense gravitational stress upon both. When water eventually condensed upon the surface of the cooling Earth, tidal ranges may have been hundreds of metres high while the Moon would have taken only a few days to orbit its partner. Today the Moon and the Earth are quite a distance apart, the highest tidal ranges amounting to a mere 2 metres in open sea and a modest 16 metres or so on some coastlines. The Moon now takes much longer, 27 ⅓ days, to orbit the Earth.

The Lunar Month

The mean Lunar Month (same phase to same phase) is 29 ½ days, slightly longer than the mean orbit of the Moon - the Earth moves part of the way along its own orbit before the Moon catches up and realigns itself with the Sun.

This circling dance around the pirouetting Earth does not keep to a rigid, mechanical beat however. It follows a more playful rhythm. Due to the Moon's elliptical orbit (see below) its motion is constantly speeding up and slowing down slightly. As a result of this the number of days that can be counted from one Newmoon until the next is either 29 (short month) or 30 (long month). These short and long lunar months were used in the calendars of many societies including the ancient Greeks, aswell as the Iron Age Druids who interpreted them as good omens (even-numbered months) or bad (odd-numbered) [Coligny Calendar, Musée de la Civilisation Gallo-Romaine, Lyon].

The Moon's Elliptical Orbit

The Moon's closest approach to Earth (perigee) on its elliptical orbit can vary from its most distant position (apogee) by as much as 10%. The imaginary axis between apogee and perigee, known as the lunar apse, itself rotates in an 8.85 year cycle. This elliptical orbit is the main element giving rise to the varying lengths of each lunar month as the Moon speeds up towards perigee and slows down towards apogee (fig 1).

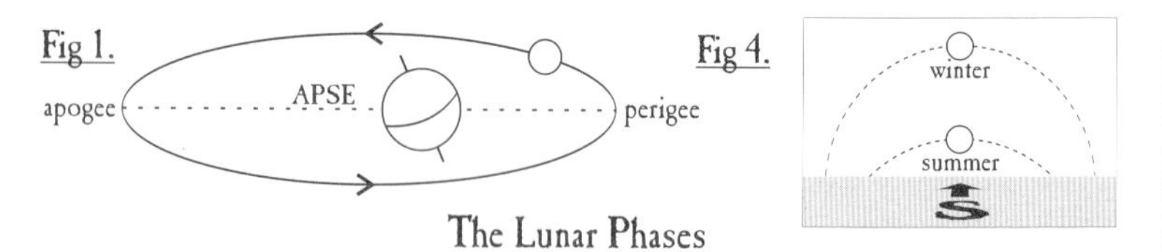

The Lunar Phases

The Sun shines its light upon the half of the Moon that faces it. Depending on where the Moon is on its orbit, the Earth sees a different view of this illuminated half. The sunshining face of the Moon appears to wax and wane as it comes into view and goes out of view again around its orbit (fig 2).

The Rising and Setting Times of the Phases

As the Moon's changing phase is due to its changing position around the Earth, so accordingly the Moon will appear to rise and set at different times of the day (fig 3).

The Winter Fullmoon

As the year passes by we see the noonday Sun appear to move higher in the sky towards summer and lower towards winter. If we watch the *Fullmoon* throughout the year we find the same happens but in exactly the opposite way to the Sun - lowest in midsummer and highest in midwinter. So, the Sun rides high in the long days of summer lighting the day; the Fullmoon rides high in the long nights of winter illuminating the night - fig 4 opp.

Fig 2.

Fullmoon
waxing Halfmoon
waning Halfmoon
Darkmoon

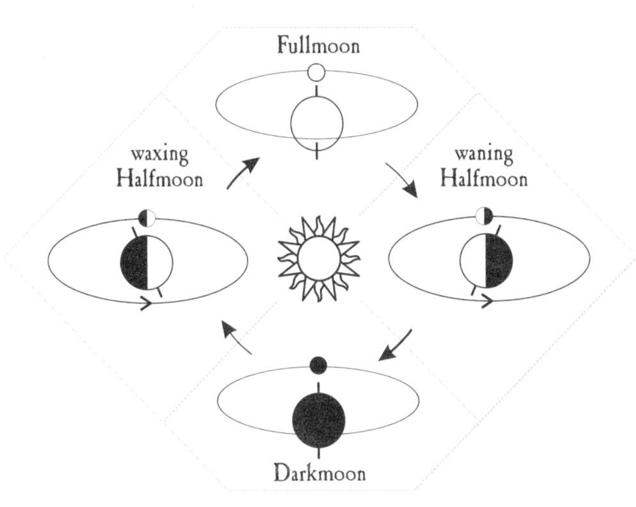

Fig 3.

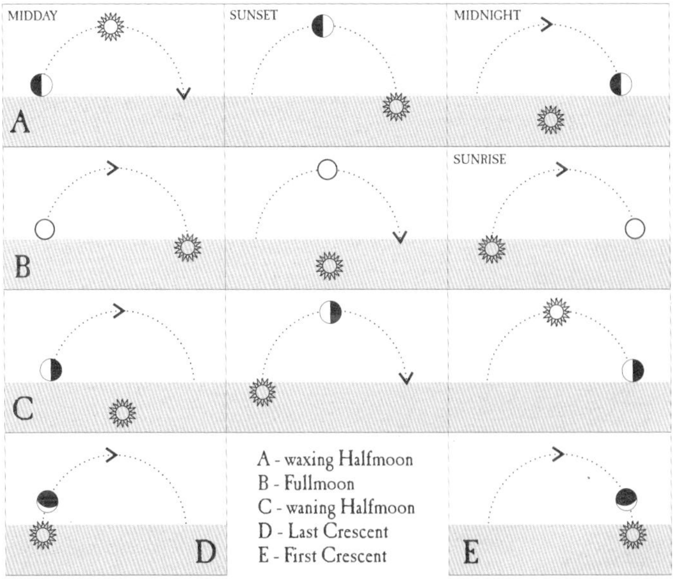

MIDDAY
SUNSET
MIDNIGHT
SUNRISE
A
B
C
D
E
A - waxing Halfmoon
B - Fullmoon
C - waning Halfmoon
D - Last Crescent
E - First Crescent

Similarly, at the time of the spring Equinox the waxing Halfmoon is at its highest in the sky and at the autumn Equinox the waning Halfmoon is at its yearly *zenith*.

The Tides

As the Earth spins around, its great fluid masses try to shift outwards in *centrifugal* motion. At the same time the Moon shifts the combined centre of gravity of the Earth & Moon (the Barycentre) away from the centre of the Earth to a position nearer the surface facing the Moon (the gravitational influence of the Sun has a similar effect, 46% that of the Moon's). This shift in the Barycentre pulls the fluid masses of the Earth towards this side creating a bulge of massed water that produces the tide, while on the other side of the Earth the resulting greater distance from the Barycentre allows the oceans to be flung further out by the centrifugal effect of the Earth's spin producing another tide of massed water. So there are two tidal bulges on the Earth, one pulled towards and one escaping away from the circling centre of attraction, which in turn leaves a lower water level on either side of these two bulges - the low tides (fig 5).

Fig 5.

⊗ Centre of Spin of the Earth (Axis)

+ Centre of Gravity of the Earth-Moon system (Barycentre)

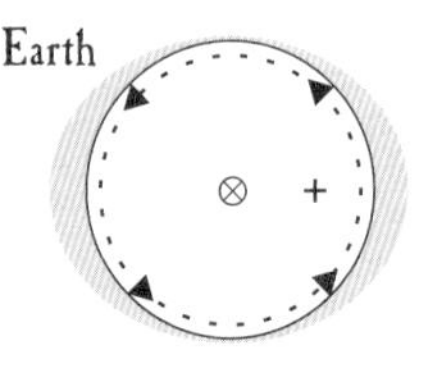

For any given place on the Earth's oceans the high and low tides will each normally occur twice a day when the Moon reaches the same positions above or below the horizon. Yet, because of the effects of coastal geography upon the tidal wave, tides can occur much later or earlier just by moving along the same coastline a little. Actual timing of tides and their range is highly dependent upon location and weather conditions as much as the position of the Moon, though the tides will occur everywhere to the lunar rhythm.

If all other things are considered equal a few basic rules of thumb can be observed underlying this more complex behaviour -

- The same tide arrives 50 ½ minutes later each day, the Moon having moved on slightly in its month long orbit around the Earth
- The high tide occurs twice each day, 12 hrs 25 mins apart. The low tide is the same but occurs 6 ¼ hrs after each high tide
- Twice a month at *syzygy* (the Full and New moons) the tides exhibit their greatest range between low and high tide (spring tides) due to the gravitational influence of both the Sun and Moon acting in alignment upon the Earth. Twice a month at the Halfmoons the tides exhibit their lowest range (neap tides) due to the countering effect of the Sun and Moon's gravity acting at right-angles with the Earth

The Eclipses & the Lunar Nodes

Most of the planets in the Solar System including the Earth orbit anticlockwise around the Sun in the same plane, known as the ecliptic. The Moon itself orbits at a slight incline to the ecliptic of about 5.15 degrees (fig 6).

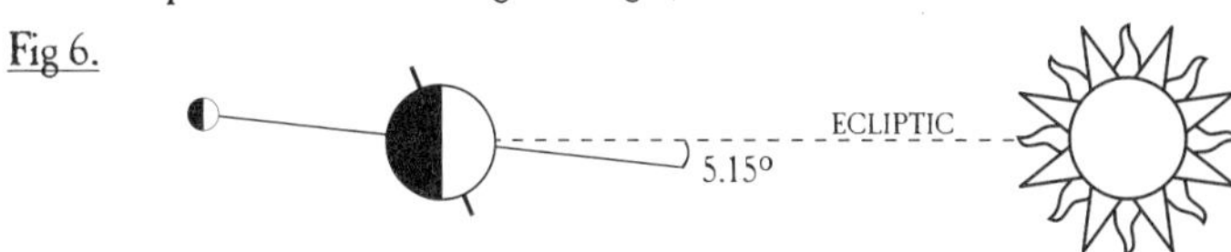

Fig 6.

The two points where the Moon's orbit cuts across the ecliptic are known as the nodes and were thought of as the head and tail of an enormous celestial Dragon. About twice a year when these nodes line up with the Sun there will usually be an eclipse or two and the Dragon is said to have swallowed the Sun or the Moon. A Solar Eclipse (either partial, annular or total) occurs if there is a New Moon on this alignment. A Lunar Eclipse (either partial, penumbral or umbral) occurs on a Fullmoon. For most of the year at syzygy the Moon does not align directly with the ecliptic which is why we do not have eclipses every month (fig 7).

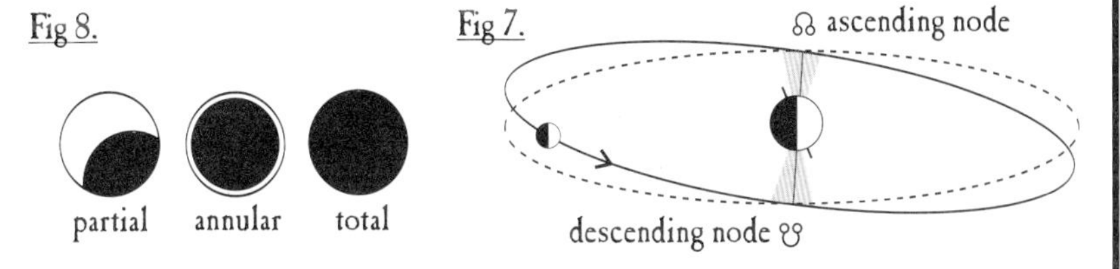

Fig 8. Fig 7.

The nodes themselves rotate clockwise over 18.6 years (known as the Draconic Year after the Dragon) so that the two eclipse seasons occur about 20 days earlier each year. Some early astronomers made simple *volvels* in the likeness of a Dragon that allowed them to predict these eclipse seasons.

The awe-inspiring Solar Eclipse reveals how the apparent sizes of the Sun and the Moon in the sky are identical, about ½° of arc. The Sun is about 400 times the diameter of the Moon and is also 400 times further away, hence today's observer sees them both as the same size in the sky. However, when the Moon is near to its closest approach to the Earth (perigee) its apparent size is slightly larger (up to 25% by area than at apogee - and up to 30% brighter at Fullmoon). As a result of this, when the Moon is precisely on a node at New Moon the Sun's disc will be totally obscured - a total eclipse. When the Moon is further away from the Earth (nearer apogee) it will not quite cover all of the Sun's disc on a node, and this gives rise to an annular eclipse (fig 8). A Lunar Eclipse on the other hand occurs when a Fullmoon passes through either the Earth's outer shadow (penumbral eclipse) or its darker, central shadow (umbral eclipse).

The Major and Minor Standstills

Throughout a single lunar month the Moon will be seen rising and setting at different positions along the horizon each day, moving between its most northerly and southerly limits. This movement in turn follows a yearly pattern and a longer term pattern of ever widening and narrowing ranges. The Major Standstill shows the widest of all, and 9.3 years later the Minor Standstill shows the narrowest of all. This 18.6 year cycle is the result of the same nodal cycle that gives rise to the eclipses. The changing altitude of the Moon in the sky that results determines the general illumination on the ground and the direction of pull on the tides.

Fig 9.

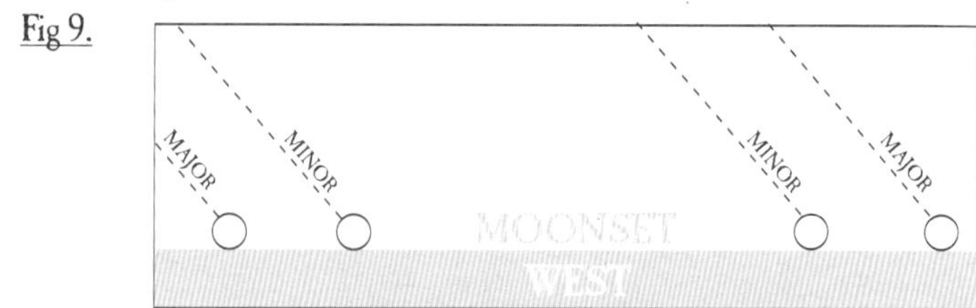

The Moon and Life

Life on Earth truly dances to the music of the spheres. Turning with the Earth, life is exposed to the daily rhythms of light and heat from the Sun, the ebb and flow of the tides, the monthly brightening and darkening of the Moon and the annual return of the seasons.

Chronobiology - the rhythms of life

Chronobiology is that branch of the biological sciences which studies the full range of the rhythmical processes of organisms - biological rhythms.

A beat is a rigid repetition that is always the same and never changes, a quality most apparent in mechanisms. A rhythm on the other hand is the repetition of similar events at similar intervals which may themselves transmute over time. Just listen carefully to some well composed music and notice how the rhythm alters over time and yet remains coherent and recognisable; all of the harmonious elements of the music are held together and immersed within these patterns of time. Similarly, organisms can be understood as being composed of multitudes of biological processes all held together within patterns of harmonious *temporal* coordination - the rhythms of life.

These biological rhythms can be visualised in a chronological scale and compared with similar environmental rhythms found in nature (fig 10).

Fig 10.

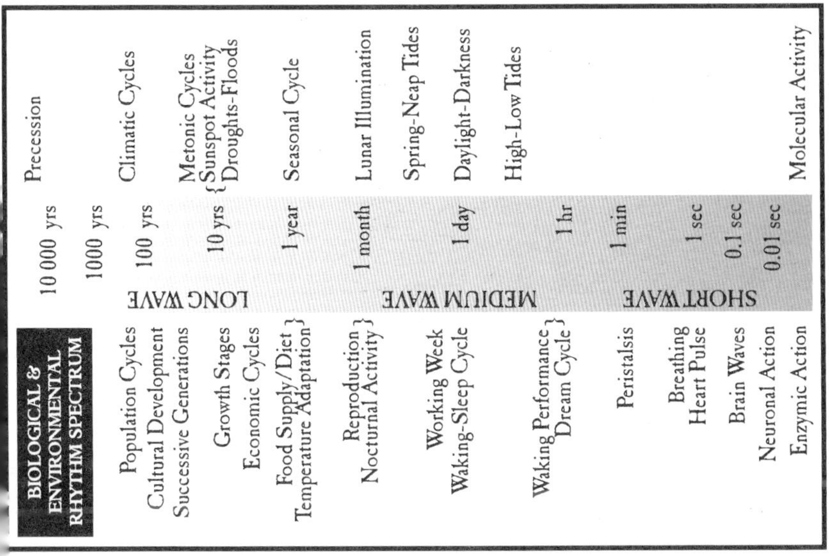

BIOLOGICAL & ENVIRONMENTAL RHYTHM SPECTRUM

Biological	Wave	Period	Environmental
		10 000 yrs	Precession
		1000 yrs	
Population Cycles Cultural Development	LONG WAVE	100 yrs	Climatic Cycles
Successive Generations			
Growth Stages		10 yrs	Metonic Cycles Sunspot Activity Droughts-Floods
Economic Cycles			
Food Supply/Diet Temperature Adaptation		1 year	Seasonal Cycle
Reproduction Nocturnal Activity		1 month	Lunar Illumination
	MEDIUM WAVE		Spring-Neap Tides
Working Week Waking-Sleep Cycle		1 day	Daylight-Darkness
			High-Low Tides
Waking Performance Dream Cycle		1 hr	
Peristalsis	SHORT WAVE	1 min	
Breathing Heart Pulse		1 sec	
Brain Waves		0.1 sec	
Neuronal Action		0.01 sec	
Enzymic Action			Molecular Activity

An external event that controls a biological rhythm, such as the seasonal fluctuations of temperature which force many animals to migrate or hibernate each year, is referred to as an Exorhythm. If the cause is an internal biological source it is referred to as an Endorhythm ('body clock'). An Exo-endorhythm on the other hand is any internal source of rhythm that requires an external source to arouse it andor synchronise with.

Chronobiological processes are essential for understanding the constantly changing patterns of life in cells, multicellular organisms, biological societies and ecosystems right up to the level of the biosphere itself (such as the effects of periodic glaciations and global warming). Nothing is static or linear; all is alive and adapting to the continuously changing complex of patterns in time, ever repeating and yet always evolving; never exactly the same and yet never entirely different. The rhythmical dance that binds all together.

Some Exorhythms

As the Earth orbits around the Sun its 'gyroscopic' axis remains pointing in the same direction in space. So as it circles around the Sun, the Earth points slowly towards and then away from the Sun giving rise to the seasonal changes in day length and temperature (fig 11). As a result theMoonbecomes an even more important source of illumination during the darker seasons when the Fullmoon rides higher in the sky.

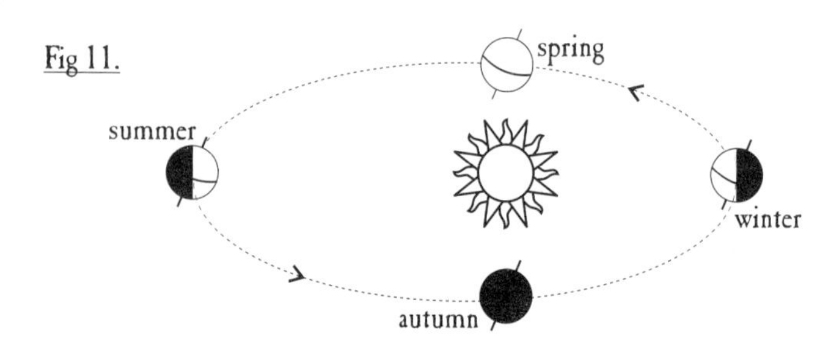

Moonlight Intensity

At full illumination the Moon's light is powerful enough to stimulate our colour vision (normally reserved for daylight activity). Indeed one particular phenomenon of the changing phases sends a powerful pulsed signal to life below every month. As the Moon waxes it brightens gradually until about a day before the Fullmoon when its brightness rises sharply by about 30%. It subsequently peaks at the precise moment of Fullmoon when the lunar light is twice as bright as it was 24 hours earlier. This lunar opposition effect diminishes in a similar manner over the following day [2]. The rapid brightening of moonlight dramatically increases night visibility at the times of the Fullmoon, turning the night on Earth into lunar twilight - ideal conditions for *mesopic* vision.

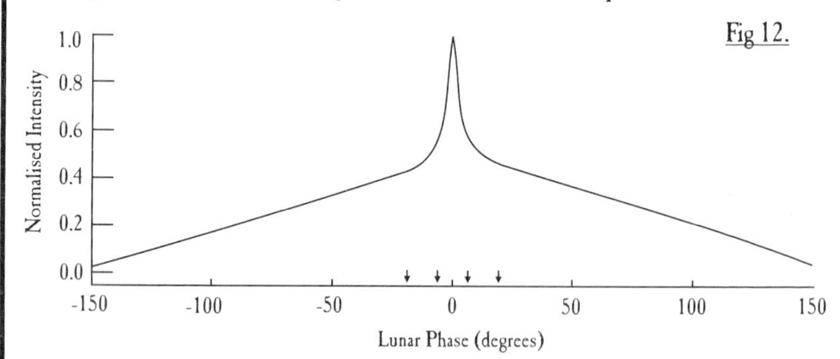

The Metonic Cycle

The same lunar phase will occur about the same day of the year every 8, 11 and 19 years. This gradual aligning and misaligning of the same lunar phase with the same day of the year is known as the Metonic Cycle and gives rise to the superimposition of two environmental cues used by organisms - the seasonal and mon[illegible] rhythms. The actual appearance of a specific phase (or its reverse phase) on or near to [illegible] day of the year has been found to coincide with patterns in th[illegible] isms (most notably with changes in popul[illegible]

Full Cycle	Lunations	Accuracy	Half Cycle†	Lunations	Accuracy
19 yrs	235	2 hrs	15 yrs	185 ½	17 hrs
11 yrs	136	36 hrs	4 yrs	49 ½	19 hrs
8 yrs	99	38 hrs	9.5 yrs‡	117 ½	1 hr
3 yrs	37	74 hrs			
9.5 yrs*	117 ½	1 hr			

*alternating 8 & 11 year Full Cycles
† alternating reverse phases
‡ alternating 4 & 15 year Half Cycles

There are many other examples of cosmic rhythms that life is exposed to, such as the 11 year sunspot cycle altering the amount of solar radiation that bathes the Earth (caused by the 22 year cycle of the reversing magnetic poles of the Sun, and particularly important for chlorophyll-containing organisms and the food-chain that feeds off them); the 41,000 year cycle of the Earth's axis moving between 22 & 24 ½ degrees inclination; the 95,000 year cycle of the Earth's orbit moving between ellipsoid and near circular; the 21,000 year *precession* of the Earth's axis relative to its orbital eccentricity determining which hemisphere of the globe faces the Sun at *perihelion*. The combination of the latter three cycles is believed to influence the rhythm of the ice ages and glacial & interglacial periods due to the effect on global temperature.

From Sea to Dry Land

Life is fundamentally attuned and adapted to these fascinating heavenly rhythms. For the first 90% of time that life has existed on Earth it has mainly inhabited and adapted to the surface waters of the oceans, constantly under the influence of the daily tides and alternating light of Sun and Moon. Only in the last 10% of this time has complex life adapted to and colonised dry land away from the oceans, under the full glare of Sun and Moon.

When much closer to the Earth, the Moon's influence on the oceans created a tidal zone on the coasts, rhythmically throwing up organisms over the shorelines. This provided the initial impetus for life to develop away from the amniotic environment of the sea and adapt to a partial exposure to dry land, gravity and the wildly fluctuating temperature of the desiccating air. This adaptation to the tidal shorelands provided complex life with the stepping stone it needed to gain a permanent foothold upon the land.

A selection of organ[illegible] displaying lunar synchronised activity, whether through tidal changes or the [illegible] of moo[illegible]ight, clearly demonstrates the importance of the lunar rhyth[illegible] some as the seasonal rhythm of the solar year [[illegible]]

The Moon and Protoctista

Many of the Protoctista (protozoa, algae, water moulds etc.) such as the luminous algae off Brittany and the diatoms off Cape Cod show a daily lunar activity, migrating to the surface with each low tide. Others such as the brown algae off Morocco, Jamaica, North Carolina, the Isle of Man, Scotland, Wales, Misaki, South Australia and the Monterey Peninsula *spawn* and reproduce on or around the times of syzygy (Fullmoon and New Moon).

The Moon and Plant life

The halodule of the Fiji Islands release their pollen into tidal pools to meet the stigmas of the female plants precisely at the times of the lowest water level of the spring tides around syzygy. The water plant enhalus of Banda Island, Indonesia, only blossoms at the syzygy spring tides, with the male blossoms rising underwater to meet with the female blossoms on the surface at low tide.

Seeds *germinated* in propagators just prior to the Fullmoon grow larger, forming larger numbers of blossoms and producing greater harvests than those sown just prior to the New Moon in the case of wheat, oats, barley, maize, peas, beans, carrots, head lettuce, white cabbage, leek, lovage, yarrow and others.

The *blossoming* period of pearl barley, common wheat, soya bean, rocket candytuft and corn cockle is shortened when exposed to moonlight. Statistically there has been a better grape harvest if a New Moon occurred in the first half of June - the time of the vine blossom (Metonic Cycle).

The humble potato absorbs more oxygen (increased metabolism) when the Moon is setting and least when the Moon is rising, with a monthly maximum at syzygy. While the seeds of the common bean absorb water at their greatest rate at syzygy and the Halfmoons, all other things being equal.

The Moon and Animals

As many prey animals swarm and breed to the cue of the Moon so their predators take advantage, with many adopting similar cycles in their behaviour.

SEA

Sponges, Coral, Marine Worms, Shell Fish

Sponges off Curaçao, the Great Barrier Reef, Okinawa, and the Firth of Clyde; corals off western Australia, the Great Barrier Reef, Puerto Rico, Panama, Costa Rica, Jamaica, Okinawa, the Marshall Islands and around Hawaii all spawn in synchrony with the lunar cycle.

Marine worms off Brittany, Normandy, the Mediterranean, the Bahamas, Bermuda, Vancouver, California, Florida, Massachusetts, Japan, Bay of Bengal, Samoa, Indonesia and many places besides all show a lunar rhythm in their reproduction at certain times of the solar year, each species *swarming* around a particularly favoured phase of the Moon.

There are many sea snails, periwinkles, cockles, mussels, oysters and scallops around the world with metabolic cycles, feeding times and spawnings all timed-in with the tides and periods of the Moon at specific seasons of the solar year.

Sea Urchins, Small Crustaceans

Sea urchins swarm off Japan, Panama, Southern California, Florida Keys, and the Red Sea reaching a maximum at New and Full moons, and many sea slugs and sea cucumbers swarm and spawn with the Moon.

Many species of shrimp deposit their eggs andor moult in phase with the lunar cycle. Many species of krill and shorehopper have reproductive, feeding and other behaviours in harmony with the tides and periods of the Moon.

Crabs & Lobsters

The atlantic crab shows a lunar rhythm in its egg-deposits with a maximum around Fullmoon. In Apalachee Bay, Florida the newly hatched larvae of the same crab move to the surface from their nests under the beach on the night of the spring tides, particularly under the Fullmoon.

The spiny lobster in the Torres Straight moults within the third quarter of the Moon's cycle. The blue crab off Maryland moults with a maximum moulting activity at Fullmoon. The larvae of shore crabs in Ria de Aveiro, Portugal, have their maximum appearance in the darkest part of the lunar cycle at low tide. Around Okinawa, Japan, a particular crab moves her sticky eggs to the back of her body at the Darkmoon for the larvae to be released two weeks later at the Fullmoon, after which she moults. A courting male crab of the canal zone of Panama puts the finishing touches to its cave at the time of the Fullmoon. The cave production of a crab in the Seychelles reaches its maximum around the Newmoon. And a terrestrial crab in Japan releases its larvae in the lower reaches of the river only at the spring tides around syzygy.

Fish

Most eels migrate to Cumbria and Norway when the high tide coincides with the dark nights of the Moon and avoid the Fullmoon. Greater catches of shortfin eel are made as they gather for migration under the waning Moon and especially at the waning Halfmoon.

Catches of the atlantic herring are greatest in the autumn and winter around Fullmoon.

In Scotland the emigration of young salmon is least around the Fullmoon. The catch rate of a lake trout in Switzerland is least around Fullmoon and greatest around Darkmoon. The masu salmon of Japan start their migration down river under the Darkmoon.

The majority of european graylings move upstream for spawning at the New Moon.

The sea bass off Papua New Guinea spawns at syzygy. From February to September on the beaches of California the grunion fish spawns en masse a few days after syzygy, on the high tide, making it a very easy catch indeed.

Many species of tropical fish spawn and feed in rhythm with the Moon. The blue grenadier of South Australia has its maximum spawning in the *last quarter* of the lunar cycle. The feeding behaviour of the butterfly fish is tied directly to the lunar spawning behaviour of the polyps it feeds off.

AIR

Insects

Flying insects in Kenya fly higher as a whole under the waxing Moon and lower at the waning. There are more moths flying at syzygy than at the Halfmoons. In an oak forest in Quebec it was found that most flying insect activity occurred between Fullmoon and waning Halfmoon under stable temperature conditions.

Many species of moth in Trinidad, the owlet moth in South Queensland, Australia and the cutworm in Varnasi, India, are caught least in light traps at Fullmoon and most at New Moon.

The bollworm moths in Texas lay their largest deposit of eggs just after the Fullmoon and smallest at the New Moon.

Mayflies in Switzerland, Punjab, Uganda and Cameroon all display a lunar rhythm in swarming.

Caddis flies in Uganda show a lunar rhythm in hatching and swarming.

Female mosquitoes are caught in light-traps more frequently in Australia and Trinidad around the Fullmoon.

Most giant water bugs in the Ivory Coast fly into light traps around the Brightmoon and least around the Darkmoon.

Light traps in India catch more rice green leafhoppers towards New Moon and less towards Fullmoon.

When the entrance of the nest of the carnica honey bee faces north the bee reaches its maximum flying activity at the Fullmoon. When the nest is aligned east-west, maximum flying activity occurs at the New Moon, throughout most of the year.

Birds

Fewer birds fly into artificially illuminated obstacles ('light traps') at night under the Fullmoon. (Many birds use the stars and Moon to navigate when migrating.)

In Finland when the Fullmoon coincides with a particular time of the year (Metonic Cycle) reproduction increases in the capercaillie and black grouse.

During the winter months lapwings in Hampshire start to engage in a nightly search for food around the Fullmoon but at no other time. In Spain the common crane goes to its sleeping-perch later and later after sunset as the Moon waxes.

In Venezuela the thick-billed plover uses the light of the Moon in the winter months to extend its hunt for food on the tidal flats during the long nights: more so at times of nightly high-water.

Every 10 lunar months the sooty tern returns to Ascension Island to nest and brood.

A nesting colony of black-naped terns on Eagle Island always laid their first and following clutches of eggs around the New Moons. Feeding activity was higher at the spring tides.

Marauding seagulls attack the cassin's auklet off California particularly around the Fullmoon.

The great horned owl and the spotted owl in the USA showed more frequent calling behaviour at specific times of the lunar month.

The nightjar in Plymouth shows a lunar rhythm of courting-song and egg laying.

Bats

Leaf-nosed bats in Panama and Columbia showed reduced activity of seeking food on nights around the Fullmoon.

LAND

Amphibians

Certain toads in Bali and Java lay their eggs in rhythm with the Moon, as do certain coastal frogs in Java. The black-spined toad for example spawns under the waxing Moon so that its offspring mature by the Fullmoon.

Reptiles

The Israeli house gecko increases its nightly activity with the waxing Moon and decreases it with the waning Moon.

Rodents

The kangaroo rat in the California deserts keeps a low profile whenever the Moon is out at night, remaining either in its daytime shelters or under cover and close to its shelter. The grasshopper mouse in Oklahoma spends more time outside its shelter around the Darkmoon, as does the wood mouse in Scotland during the winter months.

The golden hamster is most active running about just before Fullmoon.

Larger Animals

The porcupine of the Negev Desert in Israel avoids the moonlight in the autumn, winter and early spring months.

The reproductive rate of the snowshoe hare in Canada shows a mass appearance on average every 9.6 years - Metonic Cycle (based on 100 years of data). The closer the Fullmoon falls to a specific spring date the greater the rate of reproduction. The population of its predator the canadian lynx fluctuates accordingly, about a year or so behind that of the hare.

About twice as many galapagos fur seals are found on the land at Fullmoon than at Darkmoon, whereas the common seals of Snake Island in British Columbia show an increased tendency to stay in the water at Fullmoon.

The rutting of the impala seems to start around the Fullmoon. The indian buffalo mates at the New Moon and occasionally at Fullmoon

Many placental animals show gestation periods in multiples of 30 days (lunar month).

Human beings show marked behavioural patterns linked to the Moon, involving some biological synchronisations and many culturally mediated behaviours and observances (see below).

(Full details of these and many other organisms can be found in Endres & Schad, 2002 [3])

The Moon and Humans

"...man was first jolted into calculation and generalisation by the need to keep track of the Moon; that from the Moon came calendars from them, mathematics and astronomy (and religion too); and from them, everything else. As the Moon made Man possible as a physical being through its tides, it made him an intellectual being through its phases."

Isaac Asimov - The Triple Triumph of the Moon [4]

The Moon is the Measure

moon - Germanic *mond*, Greek *men*, Latin *mensis*, Indo-european root *me*, *mens* = measure

month - Greek *mene*, Latin *mensis* - the ancient and universal unit of time measured by the Moon. Deriv. - *menses; menarche; menopause; meniscus; meter; semester; symmetry* etc.

The American Heritage Dictionary of the English Language: Fourth Edition. 2000.

As we shall see, the Moon took on an altogether special significance for one creature roaming the land and shores. The hominids developed and evolved with a fascination for the Moon that would lay the foundation for Civilisation to emerge.

Human Chronobiology

Our sensitivity to colour changes throughout the year and also throughout the lunar month so that we become less sensitive towards the red end of the spectrum during summer and at New Moons, while we become more sensitive to the red end of the spectrum during winter and at Fullmoons. This is irrespective of whether we have been observing the Moon's phases or not. Similar changes in colour sensitivity have been observed in fish [3].

The normal term of a healthy human pregnancy is exactly 9 lunar months (266 days from conception; 280 days from the last menstruation).

About 28% of reproductively mature women show a 29 ½ (±1) day menstrual cycle length (lunar month). These women tend to menstruate around the Fullmoon with a diminishing likelihood of menses onset as distance from Fullmoon increases. Women whose cycles approach the 29 ½ day span have the highest likelihood of fertile cycles, while women whose cycles become longer or shorter have a proportionately diminishing incidence of fertile cycles [5].

Many women with irregular menstrual cycles who are exposed to subdued light in their bedrooms at the time of ovulation (14th day after onset of menstruation) found their menstrual cycles became regular and of about 29 days in length (lunar cycle) [6 p156].

As with the females of other animal species in oestrus, when living together, many women find their cycles become regulated by and synchronised with one another. Studies suggest this is due to the influence of *pheromones* exuded in sweat [5] detected unconsciously via the Jacobson's organ situated in the nasal septum.

The combination of the above two factors suggests that women possess an exo-endorhythm reinforced by one another and attuned to the cycles of the Moon, which is currently under-stimulated and counter-stimulated within our culture.

There is evidence that male hormonal rhythms are influenced by their female partner's cycles and vice versa. These were shown to significantly alter the timing and length of menstrual cycles [6 p278]. Women who have a regular weekly love life with a male partner tend to have significantly more 29 ½ (±3) day cycles (lunar rhythm), higher incidences of fertile patterns, and higher levels of circulating oestrogen than women who do not [5]. This suggests that human groups, male and female, evolved a rhythmical, hormonal interdependence with one another and their natural environment - bringing a whole new (or old) meaning to the term 'honeymoon'. (The Moon draws all together.)

Conditions such as Seasonal Affective Disorder (SAD), jet lag and the symptoms of sleep-depriving shift work are powerful reminders of how our physical and psychological well-being is dependent upon our exposure to and synchronisation with these natural rhythms.

Human Culture

With all the nightly activity of prey and predator affected by the tides and light of the Moon, and with the human reproductive cycle so closely linked to the Moon, it is no surprise that we find early humans and traditional societies holding the Moon in such high regard. In fact humans ordered their whole working, social and ritual timetable around the lunar cycle - hunting, farming, feasting, fertility and religion all patterned themselves with the lunar cycle as much as with the seasonal rhythm of the solar year. The human animal evolved into a truly lunar animal.

Prehistory

Before the advent of artificial lighting human activity naturally remained close to the warm hearth on the nights of the Darkmoon. With human colour vision stimulated by the light of the Fullmoon, activity would have naturally extended beyond the hearth on these nights. From early on there will have been adaptive pressure to take advantage of this cycle and to develop a monthly physiological rhythm to reflect the opportunity provided for extended hunting and foraging under the Brightmoon.

Counting Sticks

As humans sought out the order within their environment the ability to count and measure soon developed importance. One of the earliest forms of counting appears to have been accomplished by marking notches onto pieces of bone, antler or wood [7]. Many of these counting sticks, some dating back 37 000 yrs, seem to possess a distinctly lunar pattern in their counting [1 & 8]. Certainly, traditional societies such as the Bushmen of Namibia & the Sioux Indians of the American Plains used Moon-counting sticks up until the present; the Sioux carving notches for each day of the lunar month on one side, and notches for each completed lunar month on the other. This ritual object was used to keep track of the right days for conducting various ceremonies and hence to maintain the natural order between the human company and the environment they depended upon [9 p16].

Stone Circles

With the development of grain farming and animal husbandry, large-scale, permanent settlements arose. Those in charge of ritual had the time and stability required to note the movement and changes of the Moon against the solar year. The accumulated knowledge handed down over generations allowed them to recognise many diverse behaviours of the Moon which seem to have been expressed in many early structures, such as the stone circles and other monuments of Neolithic and Bronze Age Britain. The recumbent stones of the Aberdeenshire circles seem to have been arranged for the passage of the summer Fullmoon as it skirted low above the horizon in this northern latitude. Many of the ancient tombs in Ireland are aligned with major events of the solar and lunar cycles. The lightbox of a tomb at Carrowkeel in Sligo is aligned to allow the light of the Fullmoon to illuminate its chamber around the winter solstice: many of the surrounding place names have associations with the Moon [10]. The white stone and quartz which adorned many of these monuments would have gleamed brightly under the moonlight, as would have the chalk banks, slopes and moats of the great Avebury complex in Wiltshire. A sophisticated culture was emerging with a distinctly lunar flavour of observation and ritual.

Stonehenge

Above all, Stonehenge (c. 2000-1600 BCE) represents the embodiment of sophisticated lunar, astronomical knowledge marked out in stone geometry and alignments. The famous Heel stone is aligned to the Moon's rising midway between its two extreme northerly rising positions (Major and Minor Standstills). When a Fullmoon rises precisely along the Avenue the following Fullmoon will usually experience a lunar eclipse. There are 29 Z holes (short lunar month) and 30 Y holes (long month). There were 59 bluestones arranged in a circle (29 + 30) and then there were 29 ½ stones in the Sarsen Circle (59/2 - there is one Sarsen stone half the width of the others). The four Station Stones formed a rectangle the sides of which align with the Moon's most northerly setting and most southerly rising points on the horizon [11].

At Stonehenge this planetary geometry also aligns with the Sun's midsummer rising and midwinter setting. Here we find there were 19 bluestones forming a horseshoe in the centre of the complex (19 years of the Metonic Cycle) so squaring the circle of the natural solar and lunar cycles. The lunar element shines brightly throughout this highly sophisticated geometrical design.

Astronomer Priests

Before Stonehenge was completed the early priestly astronomers noted that 12 complete lunar cycles occurred within one solar year. As a result, the number 12 became a very important and key figure in early civilisation and is still used to this day in the American system of measurement. This system of numbering, known as duodecimal (base 12), is found in many civilisations throughout history, including the Romans for weights; the Babylonians, Sumerians and Assyrians for lengths, weights and time; pre-revolutionary France and the old British Imperial system for lengths and money [7 & 8].

Each lunar month the Earth moves about 1/12th of its way around the Sun so that each successive lunar month the same phase of the Moon aligns to a new position against the backdrop of stars. It was natural then for the stars of the ecliptic to be divided into 12 formal zones, the 12 stations of the Moon around the year; inspiring the 12 constellations of the celestial Zodiac; the 12 months of the year - the circle of time.

It was also noted by close observation that the Moon crossed the same background stars in the ecliptic slightly earlier in its phase-cycle each month. The Moon has a second, more hidden period known as the Sidereal Month ('star month'). There are 13 of these star-months for every 12 phase-months (to within 1 day), and hence the number 13 became associated with the Moon as its hidden number.

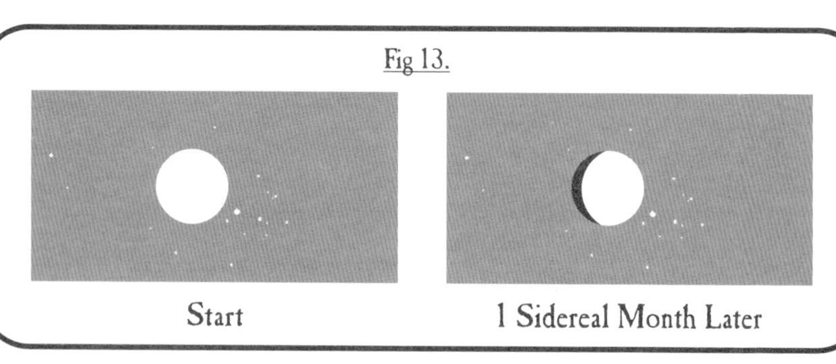

Studying the stars circling around the celestial pole the astronomer priests also noted that there were 365 whole days in the solar year: 12 lunar months contain 354 days (the so-called 'lunar year'). In many of the early counting systems these astronomer priests decided on 360 days as squaring the circle between the solar year and the 12 month lunar year, leaving 5 special, *intercalated* feast days. These 5 days were represented in the Egyptian version as the birthdays of the gods Osiris, Isis, Horus, Nephthys and Set.

12 divides 360 into 30, the approximate number of days in the lunar month. There are 60 x 6 days in the 360 day year. From this it is not hard to see how the world has come to measure time in 12/24 hours, 60 minutes/seconds, and divides the circle into 360 degrees; as indeed the foot into 12 inches, the shilling into pennies, the Zodiac into constellations, the year into months, and not forgetting the 12 dozen, the 13 'long dozen' and the gross (12 x 12).

There are 7 visible, wandering heavenly bodies: the Sun, the Moon and the 5 visible planets: Mercury, Venus, Mars, Jupiter and Saturn. Tirelessly repeating their celestial journeys these inhabitants of the starry firmament seemed to be immortal, and were considered to be gods by the early civilisations. From these 7 heavenly gods the astronomer priests derived the 7 days of our week [8].

Weekday	French	Planet	Gods: Germanic	Roman	Greek	Babylonian
Monday	Lundi	Moon		Luna	Selene	Sin
Tuesday	Mardi	Mars	Tiw	Mars	Ares	Nergal
Wednesday	Mercredi	Mercury	Woden	Mercurius	Hermes	Nabu
Thursday	Jeudi	Jupiter	Thor	Jupiter	Zeus	Marduk
Friday	Vendredi	Venus	Freya	Venus	Aphrodite	Ishtar
Saturday	Samedi	Saturn		Saturnus	Cronus	Ninurta
Sunday	Dimanche	Sun		Sol	Helius	Shamash

From the 5 digits of each hand of the astronomer priest (the 5 remaining sacred intercalated days of the year, the 5 wandering stars) was derived another system of counting - decimal (base 10).

From this heavenly order ruled by the Sun, Moon and Planets the whole *hieratic* civilisation could mimic the hierarchy of the gods in strict, professional, centralised order from Heaven down to Earth. The whole city would run as a tightly regimented imitation of the Heavenly cosmic order [12]. In the rural, agricultural world on the other hand, time remained married to the natural seasons and the lunar cycle.

In the earliest city states the most important buildings were the priest temples around which the whole city was organised. By studying the motions of these immortal wanderers, the hieratic order gained its initial impetus for the development of basic maths and geometry as well as the centralised city state with its systems of priestly accounting and geometrical architecture inspired by the order of the everlasting heavens and its immortals, the stars and planets.

The city depended upon the priestly bureaucracy to regulate and account for everything. With this knowledge of number they were able to create their own ritual calendars and keep accounts of and carefully distribute resources. They also appointed the divine warrior kings required to defend the cities when others competed for these resources. The priests had tightly ordered control over everything, just as those clockwork gods seemed to rule in the sky above - "Thy will be done on Earth as it is in Heaven."

Divorced from the natural, biological rhythm of menstruation, and the dancing rhythms of their natural environment, the insulated city folk became more and more dependent upon the artificial surrogates of these priestly intermediaries worshipping in their temples of number, in mimicry of and obedience to the heavenly order above [12].

The ascendancy of these idealised systems and their attempts to force the Moon and the Sun into marriage arrangements they could never naturally adapt to, meant that nature & female had to be repressed. The world had to give way to the artifice of the male, hieratic order - the order of the rational, numbered, counting systems - systems that would come to dominate human affairs and ultimately lead back to the Moon itself.

Religions

The lunar month is the most universal measure of all time. Most of the traditional religions of the world base their ritual year upon the lunar cycle -

Islam - entire religious calendar based upon the lunar month, each month starting from the first sighting of the First Crescent of the Moon. Ramadan is the 9th month in the cycle

Christianity - most important religious festival, Easter, and associated feast days, are determined from the first Sunday after the first Fullmoon after the spring Equinox

Judaism - lunar based religious calendar. Jewish New Year starts on the First Crescent in September or October. Passover is marked from the Fullmoon in March or April

Hinduism - months based on the lunar cycle

Buddhism - the Buddha's birthday, Wesak, is celebrated on the May Fullmoon

Chinese - calendar based on the lunar month. Chinese New Year starts on the New Moon in late January or early February

Others - Greeks, Celts, Babylonians and Sumerians all used the lunar month as the basis of their calendars

Symbolism

As the Moon wanes it seems to slowly shed its luminous skin to disappear at the Last Crescent and be reborn 3 days later as the First Crescent; as the Moon waxes it appears to grow a new skin - just like a serpent or the lining of the womb. We see the Moon and serpents often associated with female deities and fertility in world mythology.

In the Arthurian cycle there is a quest (hunt) for the Holy Grail in order to return fertility to the Kingdom. 12 shining Knights (constellations/lunar months) of the Round Table (Zodiac) and King Arthur (Sun, ruler of the Zodiac/13th sidereal month) seek to heal the barren Wasteland (infertility) by finding the castle of the Fisher King wounded in the groin (menstruation) wherein resides the Holy Grail (womb) held by the grail maiden and filled with the Holy Blood. This myth encodes the human mystery surrounding fertility and the Moon. As with the Christian mystery, 12 disciples surround the 13th Messiah King who is hung on the tree (celestial pole) to be resurrected 3 days later, and whose blood, which drips from a wound, fills the cup of the Eucharist [Acts 13:29].

The Final Frontier

It was once again the study of the Moon that provided the final crucial impetus which led to the knowledge required for the industrial age of engineering to arise. The fascination that pioneers such as Galileo, Kepler and Newton had with the Moon led directly to new laws of motion and mathematics being discovered essential for industrialisation. The clockwork universe of the astronomer priests had been resurrected.

On July 20th 1969 the whole world watched as two men landed their fragile spacecraft onto the dusty surface of the Moon. Appropriately named after the Greek sun god Apollo the 6 missions that landed a total of 12 brave men marked the culmination of thousands of years of Man's efforts to reach out and grasp the Moon. This 'giant leap for Mankind' was the natural pinnacle of the efforts of Civilisation first established by those male astronomer priests; the ultimate triumph of the arts and institutions originally derived from the very heavenly body they had now conquered with their technology.

The lunar landings also signalled a turning point in awareness as the world looked back at the pearly blue orb of the Earth floating all alone in space from the point of view of the grey, lifeless rock of the Moon. People now realise how important their natural environment is for their security and well-being, indeed survival. A signal also for a returning awareness of our natural biological rhythms and their importance to our health.

Men had finally found their way to the Moon through their priestly arts of number. Now, as women reclaim their natural relationship with the Moon, after generations of repression and separation, the two partners may marry their wisdom together for a wholly Human return to the way of the Moon.

Celestial Data

Mean Lunar Month (Synodic Month - same phase to same phase) = 29.5305888531 days (estimated 29.53058295 days in 3000 CE); min 29.274305555 days; max 29.829861111 days

Mean Sidereal Month (same star to same star) = 27.32166 days

Mean Node Cycle (Draconic Year, Minor & Major Standstills & Eclipses) = 18.618 years

Mean Solar Year (spring Equinox to spring Equinox) = 365.2424 days

Mean Orbital Precession (the rotation of Earth's tilt of axis relative to the eccentricity of Earth's orbit - influences global temperature) = est. 20,000 years (est. min 14,000 yrs; max 28,000 years; mean 21,000 yrs)

Zodiacal Precession (the rotation of Earth's tilt of axis relative to the background stars) = est. 25,770 years

Mean Lunar Diameter = 3474 km

Mean Earth Diameter = 12,735 km Current Tilt of Axis = 23° 27'

Mean Distance to Moon = 384,467 km (min 363,300 km; max 405,500 km)

(The above data have been gathered from various sources some of which may disagree slightly with the usual figures put about. This will provide some idea of how much uncertainty there is whenever absolute determination is claimed. Expert sources disagree and in reality we may only tentatively assume the certainty of the data for any time past or future.)

Lunar Terms

Lunation - the natural cycle of the Moon through all its phases and back again. On average a lunation measures 29 ½ days. In practice each lunation completes nearer to 29 days (a short month) or 30 days (a long month) - aka Lunar Month

Sidereal Month - the 27 ⅓ days, on average, taken by the Moon to orbit the Earth and realign with the same background star - aka Star Month

Waxing Moon - the former part of the month from *First Crescent* until *Fullmoon* when the Moon appears to be growing

Waning Moon - the latter part of the month from Fullmoon until the *Last Crescent* when the Moon appears to be dwindling

Darkmoon - ● the 3 days or so when the Moon is invisible to the naked eye, being obscured by the light of the Sun. The Brightmoon is the 3 days or so around Fullmoon when the moonlight is at its brightest

Newmoon - the First Crescent seen after the Darkmoon, visible in the west just after sunset. Also confusingly used to refer to the Conjunct Moon, the exact point during the Darkmoon when the Moon is in conjunction with the Sun, invisible to the naked eye (aka 'Moon conjunct Sun'). In this booklet the double-word spelling New Moon denotes the Conjunct Moon

Fullmoon - ○ the night when the Moon is at its fullest illumination and closest to due South at midnight. Astronomically the Moon is said to be in 'opposition' to the Sun - aka 'Full Moon' (Moon in opposition to the Sun)

Halfmoons - also known as the *first* and *last quarters* (see later), when the Moon appears exactly halved by light. The waxing Halfmoon (◐ *first quarter*) rises in the east about midday and the waning Halfmoon (◑ *last quarter*) rises about midnight - aka 'Half Moon' (Moon at right-angles to the Sun)

Last Crescent - the last phase of the Moon seen immediately before the Darkmoon (visible in the east just before dawn)

Young Moon - ● the phases from the First Crescent until the waxing Halfmoon

Old Moon - ● the phases from the waning Halfmoon until the Last Crescent

Gibbous Moons - the phases between the waxing Halfmoon and the Fullmoon (○ waxing Gibbous) and between the Fullmoon and the waning Halfmoon (○ waning Gibbous)

Quarters - the 4 periods of the Lunation, sometimes identified with specific phases. Strictly speaking the Moon's First Quarter is the period from the Darkmoon until the waxing Halfmoon, but just to confuse matters is also used to refer to the waxing Halfmoon itself. The Second Quarter is the period from the waxing Halfmoon until the Fullmoon. The Third Quarter is the period from the Fullmoon until the waning Halfmoon. The Last Quarter is the period from the waning Halfmoon until the Darkmoon and is also used to refer to the waning Halfmoon itself . Each Quarter lasts about a week (7 (±1) days)